Litter only

A book about dustbins
by Alexandra Martini

KÖNEMANN

Visual perception is visual thinking

There are many reasons why traveling is so great. Change, adventure, the pleasure gained from new impressions of people, culture, communication, interest in the variety of the world. Because on the spot, everything is excitingly different. This difference often manifests itself in quite banal everyday ways – it's the little difference that tells you that you're somewhere else.

So here the spotlight is on the litter bin, the trash can – often overlooked in spite of its very public appearance and general usefulness. These photographs of litter bins were taken over a period of many years, simply for the pleasure of observation, experience, documentation, and taking photographs. At the same time, they also remind me of having been in a particular place at a particular time. And to this was added the pleasure of collecting, so that today, with the contributions from family and friends, my collection contains well over 1000 photographs of litter bins in more than thirty countries on every continent.

This book is arranged by design feature, for example the type of fastening, purpose, opening, shape, volume, so the result is a kind of flipbook: the bins change in the course of the book from square to round, from top-loaders via baskets to free-standing and hanging versions.

The more love expended on litter bins, the more the container itself becomes an object worth looking at. But the minimalist version has its charms too. There is no limit to the variety of shapes, materials and colors. Viewed side by side, the objects develop a very special dynamic and reveal interesting facts about their cultural context. And so there arises a global typology of the trash can, a small litter-bin universe, which is now presented to the reader as a travel guide.

Perhaps you will be inspired by this little book to devote an un-everyday look to the everyday. Because there are surprising things to be found everywhere. And maybe the next time you are in Rio, Tokyo, Berlin or wherever, you will see among all the people a particular traffic cone or a curious sign, a colorful ice-cream van or an interesting litter bin into which you casually throw your banana skin. In any case, have fun looking, observing, photographing and bursting into the occasional broad grin.

Alexandra Martini

Some years ago Philippe Starck claimed in a BBC interview that when he comes to a new place he has not been to before, the first thing he does is go for a walk near his hotel, pick a full dustbin, carry it to his room, empty it on the bed and go through its contents in order to learn about the people, their place and the culture. Without being silly enough to check the authenticity of Starck's claim, or trying to imagine the event – it provokes some thoughts. The trash can, the final destination of unwanted objects that are now surplus and leftover.

Alexandra Martini tells us with this collection that content is not everything – look at the container, look how universal the role of the trash can is – but look at the infinite variations on the theme. This delightful album is not an academic study or a scholarly collection, it is a genuine travel diary. It leaves it to us to make our own reading of it, our own associations and understanding of the places where these vessels are designed, made, placed and used – all without having to tip their contents.

Ron Arad

Der unübliche Blick

Es gibt viele Gründe warum Reisen so schön ist. Veränderung, Abenteuerlust, die Freude an neuen Eindrücken von Menschen, Kultur, Kommunikation, das Interesse an der Vielfältigkeit der Welt. Denn vor Ort ist alles noch einmal aufregend anders. Diese Andersartigkeit manifestiert sich oft schon durch ganz banale Alltäglichkeiten, den feinen Unterschied, der spüren lässt, dass man sich bewegt hat.

So wird hier der sonst trotz seines öffentlichen Auftritts und allgemeinen Nutzens gern übersehene Mülleimer in den Mittelpunkt des Geschehens gerückt. Die Fotos von Mülleimern entstanden über viele Jahre, aus Spaß am Beobachten, Erleben, Dokumentieren und Fotografieren. Gleichzeitig erinnern sie mich auch daran, an einem bestimmten Ort zu einer bestimmten Zeit gewesen zu sein. Dazu kam die Freude am Sammeln, so dass sich heute, angereichert durch Beiträge von Freunden und Familie, weit über 1000 Fotos von Mülleimern aus über 30 Ländern aller Kontinente in meiner Sammlung befinden.

Geordnet ist dieses Buch nach gestalterischen Merkmalen, wie z. B. die Art der Befestigung, Nutzung, Öffnung, Grundform, Volumen, so dass eine Art Flipbook entsteht: Die Mülleimer wandeln sich im Verlauf des Buches von quadratischen zu runden, von Topladern über Körben zu frei stehenden und hängenden Mülleimern.

Je mehr Liebe auf den Mülleimer verwandt wird, desto mehr wird der bloße Behälter zum betrachtenswerten Gegenstand. Doch auch die Minimalversion hat ihre Reize. Der Vielfalt an Formen, Materialien und Farben sind keine Grenzen gesetzt. Einander gegenübergestellt entwickeln die Objekte eine ganz besondere Dynamik und verraten dabei auch Aufschlussreiches über den kulturellen Kontext. So ergibt sich eine weltweite Typologie des Mülleimers, ein kleines Mülleimer-Universum, für das hier nun der Reiseführer vorliegt.

Vielleicht werden Sie ja durch dieses Büchlein angeregt, das Übliche oder Alltägliche mal mit einem unüblichen Blick zu betrachten, denn überall kann Überraschendes entdeckt werden. Und womöglich steht ja, wenn Sie das nächste Mal in Rio, Tokio, Berlin oder sonst wo auf dem Globus sind, zwischen den Menschen ein besonderes Verkehrshütchen oder ein lustiges Schild, ein bunter Eisverkäuferwagen oder ein interessanter Mülleimer, in den sie eben Ihre Bananenschale werfen. Jedenfalls noch viel Spaß beim Sehen, Beobachten, Fotografieren und gelegentlichem Schmunzeln.

Alexandra Martini

Vor einigen Jahren behauptete Philippe Starck in einem BBC-Interview, dass er an einem ihm unbekannten Ort als Erstes in der Nähe seines Hotels spazieren geht, von dort einen vollen Mülleimer mit auf sein Zimmer nimmt, diesen auf seinem Bett ausleert und den Inhalt untersucht, um etwas über die Menschen, ihre Umgebung und ihre Kultur zu erfahren. Man muss nicht gleich die Authentizität von Starcks Behauptung prüfen oder sich diese Aktion vorstellen – und dennoch provoziert es unsere Gedanken. Der Abfalleimer, der Bestimmungsort ungewollter, überflüssiger und übrig gebliebener Gegenstände.

Alexandra Martini macht uns mit dieser Sammlung deutlich, dass Inhalt nicht alles ist, und präsentiert hier den Müllbehälter in seiner universalen Bedeutung und in seinen unendlichen Variationen. Dieses reizvolle Buch ist weder eine akademische Studie noch eine wissenschaftliche Sammlung – es ist ein echtes Reisetagebuch. Dabei bleibt es uns überlassen, was wir darin lesen, welche Assoziationen und welches Verständnis wir von den Orten entwickeln, für die diese Behälter gestaltet, gefertigt, positioniert und gebraucht werden – und all das, ohne ihren Inhalt auskippen zu müssen.

Ron Arad

L'autre regard

Voyager est un plaisir pour nombre de raisons : dépaysement, envie d'aventures, désir de nouvelles émotions face à des êtres et des cultures différents, besoin de communiquer, d'être curieux de l'infinie diversité du monde. Parce que tout, ailleurs, est autrement passionnant. Cette altérité se manifeste souvent au travers du quotidien le plus banal, de ce petit détail qui donne pleinement conscience d'être autre part.

Ainsi, et malgré une présence quasi officielle et d'utilité publique, voici que la poubelle, si volontiers ignorée, est au cœur de l'action. Les photographies de poubelles rassemblées ici, réalisées sur plusieurs années, sont toutes issues de mon plaisir d'observer, de participer, de documenter et de photographier. En même temps, elles me rappellent ma présence à un endroit précis et un moment précis. S'y ajoute le plaisir de collectionner qui, grâce à mes amis et à ma famille qui l'ont enrichie, me met en possession d'une galerie d'un millier de poubelles photographiées sur tous les continents dans plus de 30 pays.

Cet ouvrage s'ordonne suivant leurs caractéristiques esthétiques, par exemple la manière dont elles sont fixées, leur utilisation, leur mode d'ouverture, leur forme de base ou leur volume, pour offrir au lecteur une sorte de dépliant où, d'une page à l'autre, la poubelle se transforme de carrée en ronde, de suspendue à isolée, d'un conteneur chic en une corbeille.

Plus le regard projeté sur la poubelle sera chargé d'amour, plus ce simple récipient sera transmuté en un objet visuellement précieux. Il n'y a aucune limite à la diversité de ses formes, de ses matériaux et de ses couleurs ; jusqu'à sa version minimale qui possède d'irrésistibles attraits. Les objets en confrontation développent une dynamique particulière également révélatrice du contexte culturel. Ainsi existe-t-il une typologie universelle de la poubelle, un petit univers « poubellien » que vous retrouvez dans ce guide de voyage.

Ce petit ouvrage vous incitera peut-être à jeter un œil différent sur une banalité quotidienne où les surprises sont partout dissimulées.

Il est fort possible que lors d'un déplacement à Rio, d'un passage à Tokyo ou à Berlin, d'un voyage n'importe où de par le monde, vous puissiez apercevoir au milieu des piétons un panneau bizarre, une affiche loufoque, le triporteur bariolé d'un marchand de glace ambulant ou cette poubelle où vous jetez une peau de banane. En tous cas, toujours le plaisir de voir, d'observer, de photographier et de sourire.

Alexandra Martini

Il y a quelques années, Philippe Starck déclarait lors d'une interview à la BBC que la première chose qu'il fait lorsqu'il arrive dans une ville où il n'a jamais mis les pieds auparavant c'est, en se promenant aux alentours de son hôtel, de ramasser une poubelle pleine, de l'emporter dans sa chambre, de la vider sur son lit et de fouiller son contenu afin d'en savoir plus sur les gens, leur habitat et leur culture. Sans pousser l'absurdité jusqu'à vérifier l'authenticité des allégations de Starck ou essayer d'imaginer la scène – le geste donne matière à réflexion. La poubelle : destination dernière d'objets indésirables désormais en surplus.

Alexandra Martini nous raconte, en présentant cette collecte, que le contenu n'est pas tout et qu'il faut aussi observer le contenant, considérer l'universalité du rôle de la poubelle et envisager aussi les variantes infinies qu'offre ce thème. Ni étude théorique, ni recueil savant, ce charmant album est un véritable carnet de voyage. Il nous permet d'avoir notre lecture, notre compréhension et nos associations propres des lieux où ces récipients sont conçus, fabriqués, déposés et utilisés – sans jamais nous obliger à en déverser le contenu.

Ron Arad

Central London, UK, 5/98

Ujung Pandang, Indonesia, 4/96

Rio de Janeiro, Brazil, 12/98

CALÇADO
O
GUIMARÃES

Avenida da Liberdade, Lisbon, Portugal, 4/99

ש. וש. פּחטר בע"מ
04-8215580
24

Tel Aviv, Israel, 4/99

Vitra, Weil am Rhein, Germany, 10/97

SSI SCHÄFER

Nur Reiseabfälle!

Zuwiderhandlungen werden
als unerlaubte Sondernutzung
zur Anzeige gebracht.

Highway A9, 60km north of Munich, Germany, 12/98

Festa a SAN

Milan, Italy, 9/99

Avenida Paulista, São Paulo, Brazil, 10/99

CIDADE LIMPA
É A QUE MENOS
SE SUJA

Copacabana, Rio de Janeiro, Brazil · 10/99

Highway N4 from Johannesburg to Nelspruit, South Africa, 9/98

Eshowe, South Africa, 10/98

TOILETTE

Iguaçu, Brazil, 12/98

護美屋

Arashiyama Park Station, Kyoto, Japan, 3/99

Lisbon, Portugal, 4/99

Oslo, Norway, 6/97

Oslo, Norway, 6/97

SCORE

ADULT

GIFT SHOP

XXX VIDEOS

R50

140 SMITH STREET

Open 7 Days a Week

Durban, South Africa, 9/98

Oslo, Norway, 6/97

SPONSOR
THIS SPACE
FOR FURTHER DETAILS CALL 0171 - 361 - 5168

Exhibition Road, London, UK, 1/99

Taipei, Taiwan, 2/99

Lisbon, Portugal, 4/99

St. Pauls Ruins, Macao, China, 2/2000

KEEP DURBAN BEAUTIFUL

Durban, South Africa, 9/98

Valencia, Spain, 10/98

Denver Airport, Colorado, USA, 3/99

Tang San Street, Seoul, South Korea, 4/99

Tel Aviv, Israel, 4/99

Umahori Station, Kyoto, Japan, 3/99

RIMOZIONE FORZATA

Milan, Italy, 4/99

Niagara Falls, New York, USA, 7/95

Milan, Italy, 9/99

Wildbad Kreuth, Germany, 1/99

ゴミ入れ

Riverbed Park, Arashiyama, Kyoto, Japan, 3/99

56

เทศบาลนครเชียงใหม่

ขยะเปียก

MEDICAL CLINIC

Chiang Mai, Thailand, 2/96

LITTER

BETHNAL GREEN

1993

BROXAP

Shoreditch, London, UK, 2/98

Via Morigi, Milan, Italy, 4/99

Stansted Airport, London, UK, 4/99

Ikebukuro Station, Tokyo, Japan, 3/99

Eastbourne, UK, 5/99

Iguaçu Falls, Brazil, 12/98

あきかん

もえないゴミ
（ビン・カン）

限りある資源を大切に

Tokyo, Japan, 4/99

Barcelona Airport, Spain, 10/98

Ikebukuro, Tokyo, Japan, 2/99

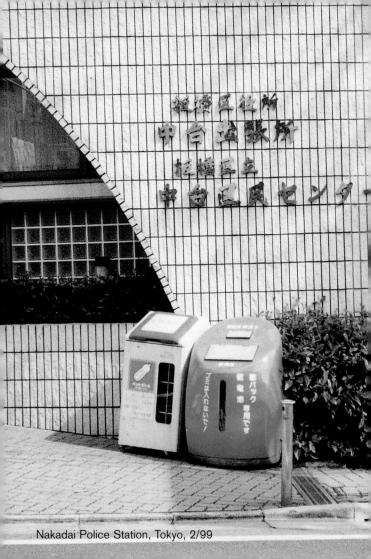

Nakadai Police Station, Tokyo, 2/99

A Pleasant Stay!

39 Bars und Restaurants, 120 Geschäfte,
13 Duty Free Shops, 2 Besucherterrassen

Frankfurt am Main Airport, Germany 4/99

Newcastle, UK, 6/98

Oslo, Norway, 6/97

Imperial College, London, UK, 1/99

LITTER

Isle of Wight, UK, 11/98

YOUR LITTER PLEASE

East Cowes Parade, Isle of Wight, UK, 11/98

BANANA
PUREE

Chiquita

Bocas del Toro, Panama, 9/95

East Village, New York, USA, 11/96

Stansted Airport, London, UK, 4/99

Studio Magistretti, Milan, Italy, 4/99

FRITUU
Meirisonne
SMOUTBOL
CROUSTILL

Antwerp, Belgium, 3/99

Deep Bay, Hong Kong, China, 2/2000

ALL RIDERS
do so at their
OWN RISK

ALL CARE BUT
NO RESPONSIBILITY
TAKEN BY
CAMEL OUTBACK SAFARIS

Enjoy Coca-Cola

Alice Springs, Northern Territory, Australia, 2/2000

Morro de São Paulo, Bahia, Brazil, 10/99.

POSTO EDEN

Lençois, Bahia, Brazil, 10/99

Highway A6 Berlin-Hannover, Germany, 8/98

Vitra, Weil am Rhein, Germany, 11/97

Royal College of Art, Kensington, London, UK, 5/99

Morro de São Paulo, Bahia, Brazil, 10/99

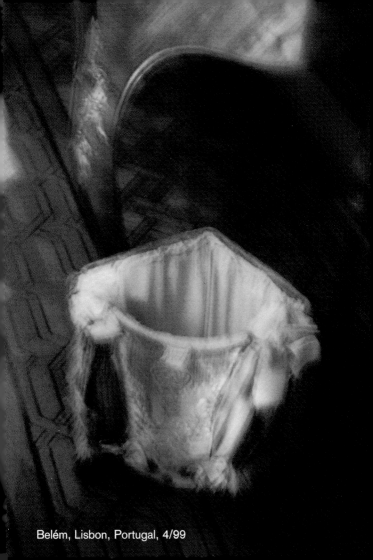

Belém, Lisbon, Portugal, 4/99

Heathrow Airport, London, UK, 12/98

I. LOVE YOU

Waterloo, London, UK, 2/99

Piccadilly Circus, Soho, London, UK, 11/99

Pedestrian Zone, Prague, Czech Republic, 3/99

Hozu River, Kyoto, Japan, 5/99

Tooting Bec Lido, London, UK, 8/99

Valencia Airport, Spain, 10/99

Iguaçu Falls, Brazil, 12/98

South Kensington, London, UK 8/98

Barajas Airport, Madrid, Spain, 10/99

Walworth Road, London, UK, 7/98

普通ごみ入れ

京都市清掃局

Side of Katsura River, "Ordinary Rubbish", Kyoto, Japan, 5/99

Waldbad Rheinau, Switzerland, 9/99

Bali, Indonesia 4/93

Bilbao, Spain, 3/99

Sugar Loaf, Rio de Janeiro, Brazil, 12/98

保持地方清潔
Please Keep Hong Kong Clean

Hong Kong, China, 2/2000

Kota Kinabalu, Malaysia, 3/99

Langa, Capetown, South Africa, 10/98

Toronto, Canada, 2/99

的士
TAXI

食物環境衛生署
FOOD AND ENVIRONMENTAL HYGIENE

Hennessy Road, Hong Kong, China, 2/2000

Máchova Ulice, Prague, Czech Republic, 3/99

Milan, Italy, 2/95

Aachenkirch, Austria, 1/99

Port Elizabeth, South Africa, 9/98

Bangkok, Thailand, 2/96

Chiang Mai, Thailand, 3/96

ขยะเปียก

456

Chiang Mai, Thailand, 3/97

Second Avenue, New York, USA, 10/99

Rio de Janeiro, Brazil, 10/99

New York, USA, 11/96

Feria, Barcelona, Spain, 2/95

East 8th Street, New York, USA, 12/96

Thompkins Square Park, New York, USA, 10/96

Fifth Avenue, New York, USA, 11/96

Castelo São Jorge, Lisbon, Portugal, 4/99

SoHo, New York, USA 11/96

Chelsea and Westminster Hospital, London, UK, 6/98

Klenzepark, Ingolstadt, Germany, 12/98

Copenhagen, Denmark, 4/99

Zeitungen, Papier
newspapers, paper

Franz-Joseph-Strauss Airport, Munich, Germany, 11/98

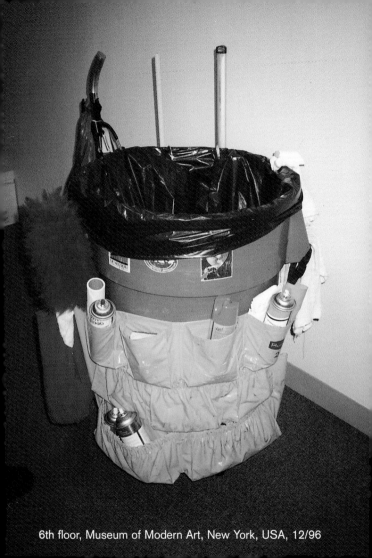

6th floor, Museum of Modern Art, New York, USA, 12/96

Thank you for your assistance in keeping these floors GUM-FREE

GUM

TARGET

New York City Transit

New York Subway Station, USA, 10/96

Helsinki, Finland, 1/99

THANK YOU
FOR NOT
LITTERING

Port Elizabeth, South Africa, 9/98

Ujung Pandang, Sulawesi, Indonesia, 4/96

Nyanga, Capetown: South Africa, 10/98

Waldbad Rheinau, Switzerland, 9/99

SMĚSNÝ ODPAD

PRAŽSKÉ SLUŽBY, a.s.
Pod šancemi 444/1, Praha 9, 180 77
Dispečink ☎ 66008666

Nusle Čestmirova Street, Prague, Czech Republic, 3/99

TEMPAT SAMPAH

Ujung Pandang, Sulawesi, Indonesia, 4/96

Bitte Abfälle hier
Pour ordures s.v.pl
Please deposit
waste here

Wildbad Kreuth, Germany, 1/99

Swellendam, South Africa, 9/98

Tel Aviv, Israel, 4/99

Abfälle
déchets
rifiuti
litter

Central Station, Zurich, Switzerland, 5/96

Swellendam, South Africa, 10/98

Technische Universität Berlin, Germany, 4/93

Hampstead Heath, North London, UK, 9/98

Rheinau, Switzerland, 9/99

Basel, Switzerland, 11/97

Bantimurung, Sulawesi, Indonesia, 4/96

Kreuzberg, Berlin, Germany, 8/98

Stavanger, Norway, 6/97

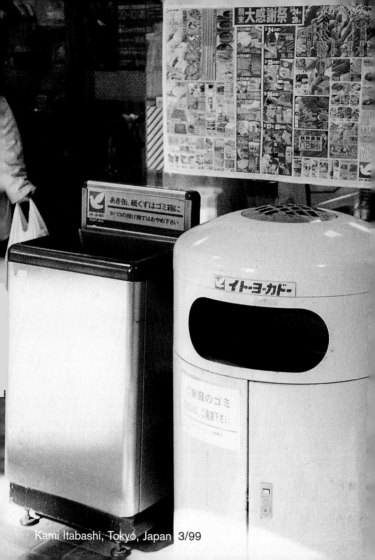

Kami Itabashi, Tokyo, Japan 3/99

凌霄閣
THE PEAK TOWER

Victoria Peak, Hong Kong, China, 2/2000

LITTER

St. Monan, Firth of Forth, UK, 4/98

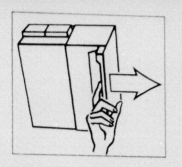

Milan, Italy, 4/99

London, UK, 4/98

Ingolstadt, Germany, 8/98

Coastal path, Kirkcaldy, UK, 4/98

St. James Park, London, UK, 6/98

Zurich, Switzerland, 6/96

South London, UK, 12/98

Skogskyrkojården, Stockholm, Sweden 3/99

Tang San Subway Station, Seoul, South Korea, 4/99

Brüssow, Uckermark, Germany, 6/97

TIEFE
1,00

Ingolstadt, Germany, 8/99

Central District, Hong Kong Island, China, 2/2000

PUBLIC.LOT. 7640

GWC

Greenwood Cemetery, Brooklyn, New York, USA, 10/96

Moabit, Berlin, Germany, 8/99

Budapest, Hungary, 7/98

Ashiya Park, Kobe, Japan, 2/99

Hochschule der Künste Berlin, Germany, 8/98

STAR GOD
OF WEALTH

財星拱照 誕生禎祥

財帛星君

吳耀慶敬

T'ien Hau Temple, Aberdeen, Hong Kong, China, 2/2000

In front of the Louvre, Paris, France, 3/97

Cascais, Portugal 4/99

Borough Tube Station, London, 3/99

Vila Madalena, Perdizes, São Paulo, Brazil, 11/99

Cascais, Portugal, 4/99

Jamaica, Queens, New York, USA, 4/95

のイヌのフンは
帰りましょう。

芦屋市・芦屋市自治環境協議会

Urake Park, Kobe, Japan, 3/99

Kameoka Park, Kyoto, Japan, 2/99

Budapest, Hungary, 10/94

Copenhagen, Denmark, 12/98

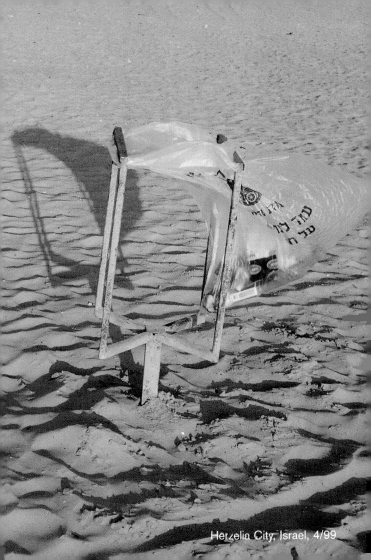

Herzelia City, Israel, 4/99

Chiang Mai, Thailand, 3/95

Tel Aviv, Israel, 4/99

Barcelona, Spain, 4/95

Mar del Plata, Argentina, 11/99

Ingolstadt, Germany, 2/99

Port Elizabeth, South Africa, 9/98

San Salvador, El Salvador, 1/99

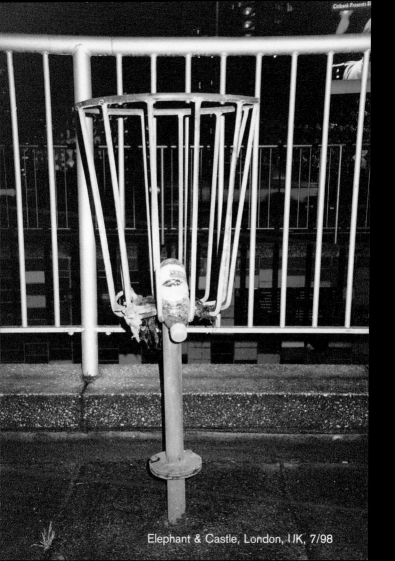

Elephant & Castle, London, UK, 7/98

San Salvador, El Salvador, 1/99

Royal Festival Pier, London, UK, 7/98

Rua Gojas, Consolaçao, São Paulo, Brazil, 10/99

Mar del Plata, Árgentina, 11/99

53rd Street, New York, USA, 9/95

Calle Consejo de Ciento, Barcelona, Spain, 2/95

New York, USA, 2/94

Taipei, Taiwan, 2/99

San Salvador, El Salvador, 2/99

Chinatown, New York, USA, 4/97

San Salvador, El Salvador, 2/99

King and Queen Street, London, UK, 4/98

Madrid, Spain, 9/99

Altpapier

Wildbad Kreuth, Germany, 1/99

ATLANTIC EXPRESS ▶

New York, USA, 11/96

Lincoln Center, New York, USA, 11/96

Barceloneta, Barcelona, Spain, 8/96

Kami Itabashi, Tokyo, Japan, 3/99

Brick Lane, East London, UK, 10/99

ЦПЯ
МУСОРА

Minsk, Belarus, 7/95

Vienna, Austria, 9/95

LA LIMPIEZA
ES SALUD

San Salvador, El Salvador, 1/99

San Salvador, El Salvador, 1/99

Macao, China, 2/2000

Não Fume

A ECONOMIA TEATRAL:
PATROCÍNIO
ou
INDEPENDÊNCIA

Centro Cultural Light

Muita atenção. Para maior
segurança e conforto dos clientes,
não é permitido:

METRÔ
RIO

R$ 1.00

FECHADO

METRÔ
RIO

Rio de Janeiro, Brazil, 10/99

Mar del Plata, Argentina, 11/99

MANTENHA
A SALA
LIMPA

Tomar, Portugal, 7/99

Uruguaiana, Rio de Janeiro, Brazil, 10/99

LITTER
and
USED TICKETS

Newcastle, UK, 6/98

Düsseldorf, Germany, 6/97

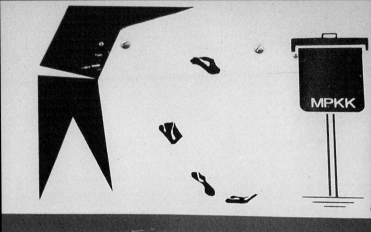

DILARANG MEMBUANG SAMPAH
NDA MAKSIMUM : RM5,0

MPKK

Kota Kinabalu, Sabah, Borneo, Malaysia, 4/99

Vila Madalena, Perdizes, São Paulo, Brazil, 10/99

Keep the Cape in Shape

Hou die Kaap in die Haak

Cape of Good Hope, South Africa, 9/98

Berlin, Germany, 8/98

Almada, Lisbon, Portugal, 7/99

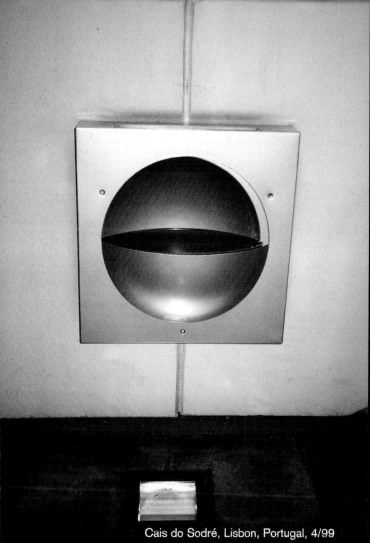

Cais do Sodré, Lisbon, Portugal, 4/99

Macao, China, 2/2000

Födermalm, Stockholm, Sweden, 4/99

Avenida Paulista, São Paulo, Brazil, 11/99

Ehrenfeldgürtel, Cologne, Germany, 2/2000

Milan, Italy, 4/99

TACHOVSKÉ NÁM.

133 207 504

DP - BUS

Žižkov, Prague, Czech Republic, 3/99

Stockholm, Sweden, 4/99

Salvador, Bahia, Brazil, 10/99

Valencia, Spain, 10/98

SCRAPHEAP
SERVICES

Tate Gallery, London, UK, 4/99

Zossener Strasse, Berlin, Germany, 1/99

Port Elizabeth, South Africa, 9/98

Aachenkirch, Austria, 1/99

Helsinki, Finland, 12/98

Hardangervidda, Norway, 6/97

Lisbon, Portugal, 4/99

Islington, London, UK, 7/98

Cascais, Portugal, 4/99

São Paulo, Brazil, 11/99

Groningen, The Netherlands, 3/99

Hampstead Heath, London, UK, 6/98

Johannesburg Airport, South Africa, 9/98

Wat Po Temple, Bangkok, Thailand, 3/95

SAC-O-MAT®

Helsinki, Finland, 12/98

robidog

Basel, Switzerland, 10/97

Marienplatz Subway Station, Munich, Germany, 2/99

San Salvador, El Salvador, 1/99

San Salvador, El Salvador, 2/99

Hluhluwe, South Africa, 9/98

Holzmarkt, Ingolstadt, Germany, 12/98

Jardines de Horta, Barcelona, Spain, 3/95

San Salvador, El Salvador, 1/99

KEEP PIGG'S PEAK CLEAN
DONATED BY

AVEX(PTY)LTD
BUILDING & CIVIL CONTRACTORS
P O BOX 3292 MANZINI

TEL 55978   FAX 54961

Manzini, Swaziland, 9/98

Kota Kinabalu, Sabah, Borneo, Malaysia, 4/99

Recoleta, Buenos Aires, Argentina, 11/99

Neukölln, Berlin, Germany, 7/96

Morro de São Paulo, Bahia, Brazil, 10/99

Johannesburg, South Africa, 9 8

San Salvador, El Salvador, 1/99

Sierra de los Padres, Mar del Plata, Argentina, 11/99

Blyde River Canyon, South Africa, 9/98

Berlin Falls, Mpumalanga, South Africa, 9/98

Royal Festival Hall, London, UK, 2/99

Valencia, Spain, 10/98

I would like here to express my particular thanks to my family as well as to my friends and acquaintances around the globe. Without the support of all of them, this book would never have become what it is now.

An dieser Stelle möchte ich meinen besonderen Dank an meine Familie sowie meine Freunde und Bekannten rund um den Globus schicken. Ohne die Unterstützung all dieser Menschen wäre das Buch nicht zu dem geworden, was es jetzt ist.

Je voudrais adresser des remerciements tout particuliers à ma famille, ainsi qu'à mes amis et à mes connaissances du monde entier. Sans leur soutien, ce livre n'aurait pas pu devenir ce qu'il est aujourd'hui.

Alexandra Martini / bin@alexandramartini.com

Concept, layout and design: Alexandra Martini
Translation into English: Michael Scuffil
Translation into French: Mariette Althaus, Françoise Chardonnier
Production: Oliver Benecke
Color separation: niemann + steggemann, Oldenburg
Printing and binding: Leo Paper Products Ltd., China
Printed in China

ISBN 3-8290-6083-1
10 9 8 7 6 5 4 3 2 1